The Dodo

Extinction in Paradise

for Rosemary

www.bunkerhillpublishing.com

First published in 2003 by Bunker Hill Publishing Inc.
26 Adams Street, Charlestown, MA 02129 USA

10 9 8 7 6 5 4 3 2 1

Library of Congress Cataloging in Publication Data available from the publisher's office

ISBN 1-59373-002-0

Designed by Louise Millar

Printed in China

The Dodo

Extinction in Paradise

ERROL FULLER

BUNKER HILL PUBLISHING
BOSTON

Dodo with Guinea Pig, *by George Edwards, c. 1758.*

Introduction

More words have been written about the dodo than about any other extinct bird. Yet the truth is that almost nothing is known of this strange creature. The bird lived only on the small, isolated island of Mauritius, way out in the Indian Ocean, and the first notice of it comes in a book published in 1599. Around 60 years later the species was extinct. What survives from this brief period of interaction with human beings is very little. There are some 15 written accounts (most of which are disappointingly lacking in informative content), a similar number of paintings (some of which contradict others), a large pile of bones, and a stuffed head and foot.

It is from this rather unpromising collection of evidence that the known facts concerning the dodo have been assembled. Literally thousands of magazine and newspaper articles, scientific papers, books, and pamphlets have been written and TV programs and films put together. Ideas and theories concerning the nature of the creature, its lifestyle, and its origins have been concocted, refuted, rejected, resurrected—mostly with only minimal evidence in support or denial. Hardly a year goes by without a new idea surfacing. Sometimes these ideas are worthwhile, sometimes they're not; sometimes their merit lies somewhere in the middle.

One might ask, Why? Why has such a mysterious creature so raised its evocative head to become one of the great icons of extinction? Perhaps the answer is simply because it is so mysterious. Perhaps it is connected to the idea that the bird came from such a romantic, tropical island. Perhaps it is due to chance and the fact that the dodo so lends itself to illustration. Clearly, the dodo's peculiar appearance—with its great beak, plump body, strange tail, and stumpy wings—is remarkable. So too is the fact that it is flightless and lacks the very quality that characterizes a bird. Then there is the name. Simple and repetitive, once heard it is unlikely to be forgotten.

Whatever the reason, one fact remains. The dodo, like the dinosaurs, has entered human awareness in a way that few creatures do. Children recognize its image in much the same way as they might recognize a lion, a tiger, a penguin, or an elephant. Yet unlike these animals, the dodo is gone and stands as a symbol for all the other creatures that have shared its fate.

The Island of Mauritius

Mauritius, home of the dodo, lies out in the Indian Ocean between Madagascar and India. It is one of three far-flung islands known as the Mascarenes. The other two are Réunion, some 100 miles (160 km) to the west, and Rodriguez, which is almost 310 miles (500 km) east. Réunion is the largest of the three, but only Mauritius has acquired any kind of renown in the outside world. The restricted fame it enjoys rests on three quite unrelated factors. A 19th-century printing error resulted in Mauritius producing some of the world's rarest and most valuable stamps; during the last few decades of the 20th century the island became famous as a vacation resort that hundreds of thousands of people visit annually; and the island was once home to the dodo.

Because of its remote location, the island now known as Mauritius seems to have remained undiscovered by humans until it was visited by Arab navigators between the 7th and 10th centuries. These early explorers seem to have made no permanent settlement. Nor did the Portuguese mariners who chanced upon the island at the start of the 16th century. Although these Portuguese sailors visited several

The Dutch fleet that visited Mauritius in 1598, painted on its return to Holland. Oil painting by Hendrik Cornelisz Vroom.

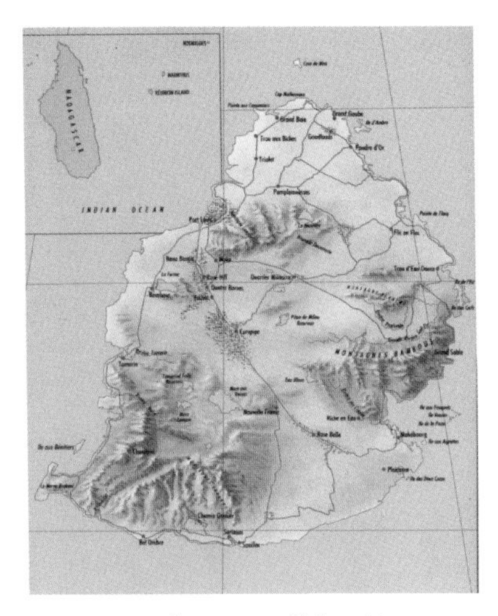

A modern map of Mauritius.

times, they were secretive about their activities, and there are no real records of what they did or how long they stayed. Proper records only start with the coming of the Dutch in 1598. On September 17 of that year several Dutch vessels under the command of Vice-Admiral Wybrandt van Warwijck sailed into a natural harbor at the southeast corner of the island, and the written history of Mauritius began.

Van Warwijck and those under his command found a virtually untouched tropical paradise. Terrestrial mammals had, of course, been unable to reach such a remote island, so wary birds hopped here and there. Giant tortoises provided a seemingly endless supply of food. Dense stands of valuable hardwoods confronted them. And, of course, there were dodos.

Unlike those who had come before them, the Dutch stayed. They were followed by the English and the French. Within a few decades the island was overrun with introduced animals, large areas of forest had been logged, and many species of native birds were vanishing.

Today Mauritius bears little resemblance to the original land of the dodo. The original forests are gone, so too are most of the birds. And, of course, the dodo is lost.

An aerial view of Mauritius.

The Dodo. *Oil on canvas by Johannes Savery, 1650.*

The Name

The word "dodo" has passed into the English language as a synonym for extinction, and the phrase "dead as a dodo" is instantly recognized by almost everyone. Short yet repetitive, it is a word that cannot easily be confused with any other. Yet the origin of the word itself is obscure.

No one can be certain whether it derives from Portuguese, from Dutch, or from some other language. One school of thought even inclines to the idea that it is simply a phonetic rendering of the bird's call. And so it might be. A two-syllable sound of this kind would certainly represent a believable bird call. Many bird species (for instance, the cuckoo) have easily recognizable calls of two syllables. However, there is no direct evidence concerning the song of the dodo—no one ever described it.

One of the most commonly held beliefs is that the word "dodo" derives from an old Portuguese word, *duodo* or *doido*, which, apparently, meant "idiot." If there is any truth in such an origin, it presumably relates to the idea that the birds seemed particularly stupid to 16th-century mariners. Alternatively, it may relate to the dodo's rather outlandish appearance.

Certainly one of the possible Dutch origins of the word relates to the dodo's appearance. The word *dodaersen* has been roughly translated as "fat behind"—a description that is fairly self-explanatory.

Another ancient Dutch word, *dronte*, was sometimes used in old texts to identify the dodo. This word is thought to have meant "swollen." Whether the word has any actual connection with "dodo" is by no means certain.

Yet another old Dutch word from which "dodo" may be derived is *dodoor*, which seems to mean "lazy" or "sluggish." One curious fact (bearing in mind the symbolic meaning that "dodo" has acquired in the English language) is that *dood*—a Dutch word for "dead"—is an anagram of "dodo."

The scientific name for the dodo is often given as *Didus ineptus*, "didus" being simply a Latinization of "dodo." Sadly, perhaps, this rather romantic technical label is no longer considered valid. According to the rules of zoological nomenclature another, more prosaic, name—*Raphus cucullatus*—has priority, and this is the rather inappropriate description by which ornithologists identify the species today.

Origins

Dodos were members of the pigeon family. Surprising though this may seem, there is no doubt that it is true. Yet dodos bear almost no visual similarity to the familiar pigeon. There are many different pigeon species in the world today, and, apart from color variation and certain differences in size, they all look remarkably alike—and in general they show not the slightest resemblance to a dodo. Only one species, the tooth-billed pigeon of Samoa (*Didunculus strigirostris*), shows the remotest of connections. There is a vague echo of the dodo's bill in the heavy beak of the tooth-bill, but even this connection is rather tenuous, for tooth-billed pigeons are small birds that can fly.

How, then, can it be said that dodos are pigeons? The answer is, because of internal anatomical similarities. The first person to notice these was a rather obscure Danish professor of zoology, J. T. Reinhardt, who, during the early 1840s, made observations on a skull that happened to be in the Royal Museum, Copenhagen. He was followed in his belief by Hugh Edwin Strickland, the co-author of the first book devoted to the dodo, *The Dodo and Its Kindred* (1848).

J. T. Reinhardt (1816–82).

H. E. Strickland (1811–53).

During the period when Reinhardt and Strickland were working, their opinion aroused all kinds of controversy. Various theories had been expressed concerning the species' similarities to other birds, and among the ideas then in vogue were connections with the vultures, the ostriches, or the bustards. The whole notion of any relationship with the familiar pigeon seemed ridiculous. Yet the anatomical proofs were, ultimately, overwhelmingly convincing, and for well over a century the link between dodos and pigeons has been accepted by scientists. Recent DNA analysis has provided additional confirmation of this strange connection.

How could a creature as large and bizarre as a dodo be descended from something that probably looked much like an ordinary pigeon? The explanation is surprisingly simple.

Two images showing how species can evolve into fatter forms when they become flightless. (Right) The swamp hen. (Left) The much stouter New Zealand takahe. Both images by Ray Harris-Ching.

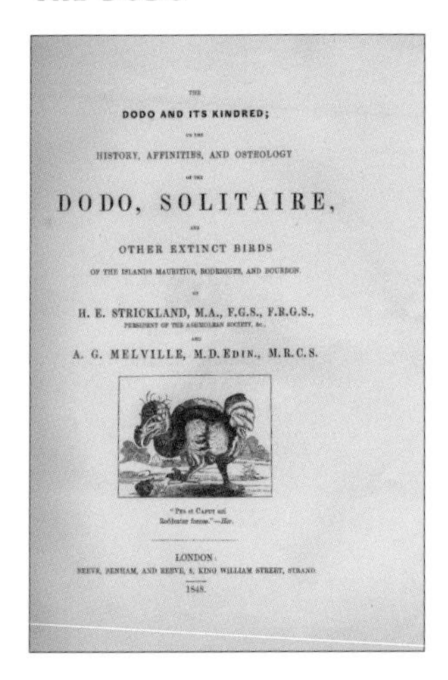

THE

DODO AND ITS KINDRED;

HISTORY, AFFINITIES, AND OSTEOLOGY

DODO, SOLITAIRE,

OTHER EXTINCT BIRDS

OF THE ISLANDS MAURITIUS, RODRIGUEZ, AND BOURBON.

H. E. STRICKLAND, M.A., F.G.S., F.R.G.S.,

PRESIDENT OF THE ASHMOLEAN SOCIETY, &c.

A. G. MELVILLE, M.D. EDIN., M.R.C.S.

LONDON:

REEVE, BENHAM, AND REEVE, 8, KING WILLIAM STREET, STRAND.

1848.

The title page of Strickland and Melville's dodo book, 1848.

By their very nature, pigeons have a great tendency to "dispersal." In other words, they have a natural inclination to make epic journeys in search of new homes. In this way they have colonized many far-flung oceanic islands, islands that most other bird kinds never reach.

Eons ago, a group of pigeonlike birds chanced on the island we now know as Mauritius. Perhaps they were blown off course by storms, perhaps they were just "wandering." The reasons don't matter. What does matter is the fact that they landed—and stayed! Why did they stay? Because they'd accidentally happened on a paradise for birds. Not only was there an abundant supply of food, but, equally important, there were no terrestrial mammalian predators—no rats, no cats, no dogs, no hogs—to chase and harry them. Such creatures were simply unable to reach a remote, oceanic island.

The Dodo. *Oil on panel by Roelandt Savery, 1626.*

Two pictures of the tooth-billed pigeon—one (color) by James Erxleben, the other by an unknown artist.

When birds reach this kind of sanctuary, certain things begin to happen. With nothing to fear, some of them lose their natural wariness and, over generations, begin to show a reluctance to fly. As the centuries and millennia pass, this reluctance increases and eventually becomes something more. It becomes an actual inability, and the descendants of typical flying birds become entirely flightless. Once a bird no longer needs to take to the air, its natural lightness (something that is, of course, essential for flight) is no longer an advantage. In fact, it generally becomes advantageous to be larger and more robust. With the passing generations individuals tend to become bulkier, with sturdier, more powerful legs. The head too can become heavier and more massive. The tail becomes shorter, and the wings (now useless) begin to atrophy.

As it has been with many birds evolving on quiet island backwaters, so it was with dodos. First they lost the need to fly, then the desire, then the ability. They got bigger, fatter, and weightier, gradually acquiring physical features to suit their new lifestyles. In short, dodos ended up looking nothing like the creatures from which they had descended.

Natural History of the Dodo

What dodos did when they were alive—how they conducted themselves, the things they ate, the sorts of places they preferred to live—is something of a mystery. The plain fact is that very, very little is actually known about the living bird.

There is only a tiny amount of primary evidence. People who actually saw the bird in life wrote only in the vaguest terms, and all that has been written or said since the time of the species' extinction is simply guesswork. Some of the guesses are probably good ones, others are probably not so good. The things that can be said with confidence, therefore, are few.

One thing that can be safely stated is that this was undoubtedly a heavy, perhaps even ponderous, bird that—because it was incapable of flight or any kind of dextrous climbing—obtained its food by foraging on the ground. What this food might have been is a matter for speculation. The likelihood is that these birds ate fruits, roots, nuts, and any other suitable vegetable matter that was

Dodo Glade. *Acrylic by Julian Hume.*

Engraving of the Mauritius seashore, 1606.

pick their way across stones and between boulders, trees, and bushes with considerable agility.

How they brought up their young, like other aspects of the birds' lifestyle, is something about which virtually nothing is known. Did dodos make nests? Did they lay one egg, or six? Did they live together in large or small groups, and did their babies form an integral part of any such groups? These are all matters about which there is little or no information

Dodos, *by S. Damrikan, from the book* Heureux Dodo.

The Coming of the Dutch, *by S. Damrikan, from the book* Heureux Dodo.

Engraving by Carolus Clusius, 1606.

One early writer, Françoise Cauche, maintained that dodos laid a single white egg in a nest of grass—but no other writer mentions this, and considerable doubt has been expressed about its authenticity. There is even doubt as to whether Cauche actually visited Mauritius. An alleged dodo egg that exists in the collection of the museum at East London, South Africa, is almost certainly the egg of an ostrich.

Were dodos aggressive creatures, or were they peaceful and timid? No one can say, but their size and formidable beak would certainly have made them among the most powerful of the original inhabitants of Mauritius. Nor is there any way of knowing how individuals interacted with other species on the island.

As far as habitat preference is concerned, the small amount of evidence that exists suggests that this was a creature that lived at, or near, the coast. This is the kind of environment in which 17th-century sailors found it. Dodos may, of course, have also lived in the interior of Mauritius, but there is no positive evidence that they did. In fact, the actual sites from which dodos were recorded are very few. This may simply reflect the kind of places that mariners actually visited rather than giving a true picture of the territory that dodos inhabited. On the other hand, it may show that dodos were very restricted in the kind of terrain that they needed.

Being large birds, dodos were probably long-lived. They seem to have been quite hardy and robust. Most dodo historians agree that the available evidence suggests that several individuals survived the long sea journey back to Europe. Only comparatively sturdy creatures could have coped with the rigors of 17th-century sailing vessels.

A commemorative dodo stamp.

The Beak of the Dodo

Perhaps the most obvious and striking dodo feature is the beak. This enormous hooked appendage, matched with the naked face, has been an object of some wonder since the day when dodos were first discovered. Its great size and peculiarity added to the mystery of the creature's relatives, and, understandably perhaps, its gross shape led many early zoologists to suppose that the dodo was nothing more than an aberrant, flightless vulture. Certainly, the beak provided no obvious clue or indication of the fact that dodos were, in reality, gigantic relatives of the pigeons.

Although the beak is such a striking feature, its precise purpose remains a mystery. Curiously, none of the writers who actually saw living dodos and described them had much to say about it–and none gives any real clue as to how it was used. Obviously, dodos used their beak for a very definite purpose, or perhaps series of purposes, but we don't know what this might have been.

Such a powerful bill could be used as a

Dodo and Shells, *by G. C. d'Hoendecoeter, 1627.*

formidable weapon—either for defense against other species or as a means by which individual dodos established and maintained some sort of "pecking order" within a group. Perhaps the beak had a function in some kind of sexual display. Perhaps it was simply a tool for feeding, enabling dodos to exploit a food resource that other creatures couldn't. Such an instrument might facilitate the breaking open of hard shells or the tearing apart of tough roots and tubers.

A naked face on a bird is usually an indicator of some kind of messy feeding activity, and so dodos may have tackled foodstuffs that would be very alien to most pigeons. They may even have scavenged carcasses washed up on the seashore.

Ultimately, all that can be said is that there is no 17th-century account to tell us what dodos did with their bills. The beak of the dodo remains an enigma.

Pen-and-ink study by Adrienne van der Venne, 1626.

Fat Dodos or Thin Ones?

Because of the doubts that surround the precise appearance of the dodo in life (due largely to the contradictions shown in the pictures produced when dodos were still living), various theories have emerged concerning its bulkiness.

Several of the surviving pictures (including the most famous, painted by Roelandt Savery) show a remarkably swollen and gross creature. Others seem to indicate an altogether sleeker bird.

Dodologists have argued over the merits of these alternatives and put forward theories and counter-theories. Since the evidence of the pictures is inconclusive, some zoologists have turned to studies of the skeleton in attempts to determine the truth. Evaluations of the load-bearing strength of the bones have been made and conjectural models constructed to show how a slimmed-down dodo might have looked. Such studies may or may not be accurate.

As with so many dodo enigmas, we will probably never know the truth. It seems likely that dodos were indeed somewhat slimmer birds than is suggested by the more extreme of the 17th-century pictures—but this is only conjecture. But whether fatter or thinner than generally imagined, there is no doubt that dodos were large and heavy birds. All serious commentators agree on this.

A slim dodo made for a display at the National Museum of Scotland, Edinburgh.

Extinction

The dodo interacted with human beings for only a very short period of time—perhaps no more than 40 or 50 years.

Presumably, the Arabian explorers who chanced upon the island now known as Mauritius around a thousand years ago saw the birds, but their records make no mention of it. Nor do the records of the Portuguese who visited Mauritius several times during the 16th century. These sailors may indeed have given the dodo its name, but no one has ever found a reference to the species in Portuguese ships' logs or journals.

It was not until the coming of the Dutch in September 1598 that any note of the dodo was set down for posterity. After their initial visit and observation, Dutch mariners came to the island again and again. They were swiftly followed by the British and the French, and for a few decades written memos concern-

Above and opposite: Two artists' impressions, c. 1890, of how a dodo hunt might have appeared.

ing the dodo occurred every now and then. Quite suddenly, toward the end of the 1630s, these stopped—and it can therefore be assumed that by 1640 the dodo was a spent force. It may not have been entirely extinct—probably a few individuals lingered here and there—but the species was so rare or localized that it was no longer being noticed.

After a gap in the written records of more than 20 years, one last description of the dodo appears, and this relates to the year 1662. This description comes in the account of a certain Volkert Evertszen, who claimed that he saw dodos on an island to which he waded at low tide from mainland Mauritius. There are some reasons for supposing that the birds Evertszen saw were not actually dodos, but most researchers believe the record to be genuine.

There is an even later written note, which occurs in a list of the products of Mauritius made by one Benjamin Harry in 1681. This list seems to be nothing more than a listing of the goods that Mauritius was famous for, and as such, the reference to the dodo was probably "historical" even at the time of Harry's writing.

To all intents and purposes the dodo was gone by the middle of the 17th century—even if a few individuals managed to survive for a decade or so longer.

Extinction, therefore, was accomplished in a remarkably short period of time. The culprit was, of course, man. Although we have no way of knowing how the dodo was faring before the arrival of the Dutch (it may, after all, have been a localized species with very precise habitat requirements), there is no real reason to suppose it was in serious decline. Presumably, dodos were well adapted to the prevailing conditions, and it was only when man altered those conditions that disaster came.

Whenever and wherever man happened upon a new land, his chief interest in the wildlife centered around the twin ideas of capture and feeding. So it was with Mauritius and the dodo. The Dutch sailors who reached the island were hungry men, sick to the teeth of the rations meted out to them aboard ship. They saw dodos, caught them–and ate them!

By and large, dodo meat does not seem to have been a hugely popular treat. Several culinary descriptions suggest that the flavor was not good. Against this it can be pointed out that hungry sailors were not particularly fussy, and dodos were, presumably, easy to spot and, being flightless, easy to hunt.

A bronze dodo cast by Pangolin Editions.

Just how vigorously people hunted dodos is not known, but they did something else that was every bit as damaging. They introduced terrestrial mammals to the island. Dogs, cats, pigs, and monkeys were released. Rats escaped from visiting ships and were introduced accidentally. Not only did such creatures compete with dodos for food and territory, they also persecuted the birds. Whether they were able to threaten the adults is a matter for speculation. However, it seems virtually certain that these creatures could destroy the young and steal the eggs. All ground-nesting birds are, of course, especially vulnerable to predation, but the dodo was confronted by an array of enemies it had never encountered before. History shows that it was helpless before such an onslaught.

A late 20th-century lead model showing a persecuted dodo.

The Dodo and Man

At first glance it may seem that there is a wealth of information concerning the living dodo, but in reality the hard evidence is tantalizingly slight and often contradictory. During the centuries since the species' extinction much has been written on the subject. There are numerous books, articles, and learned papers on every imaginable aspect, and every year brings more. But most of what is written is pure speculation. Genuine facts about the dodo are hard to come by and can be distilled from evidence of three different kinds.

First, there are physical remains. Such specimens consist largely of bone material, and several complete, or almost complete, skeletons have been constructed. It is on these that almost all serious anatomical studies are based. There are no genuine stuffed dodos to show what dodos may have looked like in life. Those that are sometimes on display in natural history museums are simply models made from the feathers of other birds.

The second kind of evidence comes in the form of paintings and drawings. Over the centuries many thousands of dodo pictures have been produced, of course, but it is only those few made during the period in which man interacted with dodos (the first 40 years or so of the 17th century) that can be used as positive indicators of the dodo's actual appearance in life. All later images are based on these and are either direct copies, copies of copies, or fanciful works that have these original contemporary pictures as their starting point. As might be expected, the contemporary pictures are rare. Some were produced by artists of considerable accomplishment; others were drawn by men with only rudimentary skills. One of the most unfortunate aspects of these pictures is that some of them tell contradictory stories, so, despite their existence, they do not enable us to be entirely sure of the dodo's typical appearance in life.

The third type of evidence lies in the writings of those who actually saw living birds. Unfortunately, there are few genuine accounts of the dodo in life, and most of those that do exist are more concerned with the dodo's suitability for the table than with any detailed explanation of its habits.

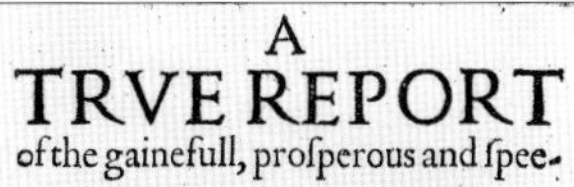

A
TRVE REPORT
of the gainefull, proſperous and ſpee-
dy voiage to Iaua in the Eaſt Indies,
performedby a fleete of eight
ſhips of Amſterdam:

WHICH SET FORTH FROM
Texell in Holland, the firſt of Maie
1598. Stilo Nouo. Whereof foure returned againe the
19. of July Anno 1599. in leſſe then 15. moneths,
the other foure went forward from
Iaua for the Moluccas.

AT LONDON
Printed by P. S. for W. Aſpley, and are to
be ſold at the ſigne of the Tygers head
in Paules Church-yard.

Title page of the book that contains the first written reference to the dodo, 1599.

Dodo bones—a 19th-century lithograph.

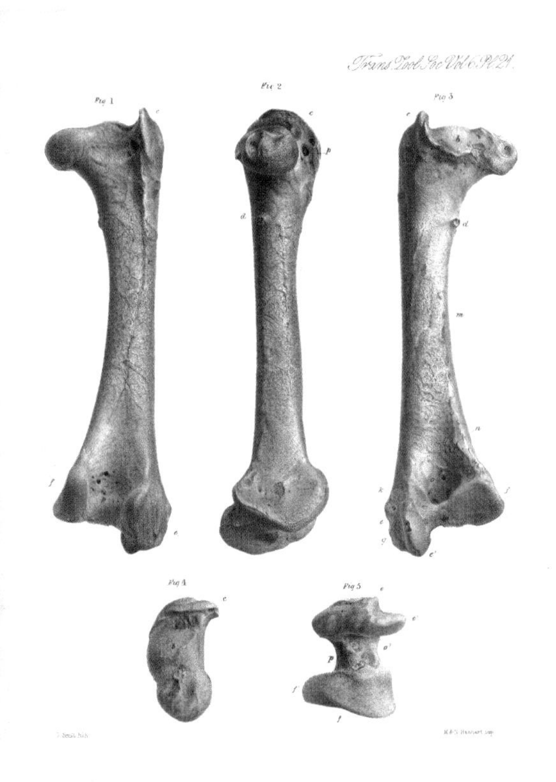

Paintings

There are some 15 paintings and drawings that constitute primary evidence for the former existence and appearance of the dodo. These are all pictures that were painted during the period when people and dodos interacted—in other words, the years between 1598 and 1640.

There are, of course, many later dodo pictures, but they are all either fanciful or heavily based on the earlier originals.

One of the curious things about the "original" pictures is that they tell somewhat conflicting stories. Some show a very fat bird, others show a slimmer one; some show physical features (markings on the beak, peculiar arrangements of the tail) that others do not; some show wings that contrast in color with contemporary written descriptions.

Dodo head, by Cornelius Saftleven, 1638.

Much has been written about these discrepancies, but what do they really mean? Unfortunately, we have no way of judging the intentions of the artists who produced the pictures. Did they care about absolute "truth to nature"? Perhaps. Perhaps not. Nor is there any real way of judging what, exactly, some of these artists were working from—living birds, dead ones, stuffed specimens, or memory. The discrepancies may relate to age, sex, or season. Sadly, there is simply not enough evidence for meaningful conclusions to be reached.

Perhaps the two best dodo pictures are also the two most enigmatic. Both show just the head, so the viewer is left to guess at the shape and proportions of the body.

The first of these images is a drawing produced by an anonymous artist who was visiting Mauritius, probably during 1601. It shows the head of a dodo that has, clearly, just been killed. The technical skill of the artist—who was, most probably, Dutch—is apparent, and it is obvious that the accuracy of the picture can be relied upon. But, strangely, it contradicts a painting produced during 1638 by an equally skilled artist, Cornelius Saftleven. Saftleven also painted just the head of a dodo—but his picture shows a bird that is very much alive. Its jaunty pose and expression indicate that Saftleven was looking at a living creature when he made his picture. Yet this is not the only way in which his image is at variance with the earlier one. He shows a dodo that has a clear patterning to the beak instead of the entirely plain one depicted by the anonymous artist. What does this mean? Perhaps we will never know.

Head of a dead dodo drawn in Mauritius by an anonymous artist, 1601.

Roelandt Savery and the Dodo

Almost every popular image of the dodo derives from the paintings of a rather obscure Dutch painter named Roelandt Savery (1576–1639). The designs so often seen on dishtowels, ashtrays, figurines—in fact, every kind of commemorative and ornamental ware—almost all have their origins in Savery's pictures.

During the first decades of the 17th century, and culminating for some now obscure reason in the year 1626, Savery made a series of paintings that included dodos. The majority of these paintings were not primarily designed to feature the species. In fact, dodos are for the most part almost incidental to the main thrust of each picture. The paintings typically show a landscape or scene with a seemingly random assortment of animals and birds wandering around within it. There is no real rationale or direction to the paintings other than that they seem to represent decorative devices designed to please an audience with a taste for the exotic. Birds from various parts of

Left and opposite: Two dodo pictures by Roelandt Savery, 1626.

the world are brought together and shown as if they form part of an ornamental garden or, perhaps, a "peaceable kingdom." Clearly, these artistic "potboilers" were bread and butter to Savery. The dodos are, in general, painted in a manner that renders them neither more nor less important than any of the other creatures that accompany them. Yet Savery chose to include dodos in his fanciful landscapes quite regularly.

Sometimes the image is an "original" one, sometimes the pose is entirely or par-

Landscape with Birds, *by Roelandt Savery, 1628.*

tially copied from an earlier picture. In general the postures are rather repetitious. There is only one known Savery painting that actually features a dodo as the main object of interest. This is an oil in the collection of the Natural History Museum, London, and now usually known as *George Edwards'* Dodo after another bird painter, George Edwards, who once owned the picture and eventually presented it to the British Museum. It is this single painting that serves as the starting point for almost all subsequent dodo images.

It is not known if Savery made a habit of including dodos in his paintings because he found them particularly fascinating, or because his customers did. Nor is it clear what Savery was actually working from. Did he have access to a living creature, or was he using a crudely stuffed bird as a model? Similarly, there is no way of knowing how realistically he portrayed the dodo. He may have been striving more for a decorative result than for a strictly accurate one. Most of his images are fairly static, but a few are much more animated and give the impression that the artist may once have seen a living—albeit captive—dodo.

Savery's dodo paintings are by no means common, but hitherto unrecorded ones do turn up from time to time, most of these showing images copied from previously known ones. Although some ornithologists seem to think otherwise, Savery was not a hugely talented painter. He was simply an artist who could produce an adequate, reasonably skillful job and occasionally manage to create images of considerable charm. In terms of the history of 17th-century Dutch and Flemish painting, he represents little more than a footnote, and it is only his dodo pictures that give him any real claim to fame. Although he achieved a certain amount of success in his lifetime, his star waned and he died in the madhouse at Utrecht in 1639.

Roelandt Savery (1576–1639).

What is absolutely certain, however, is that the dodo images created by Savery are by far the most influential of all the pictures of the species. It is the creature as painted by Savery that is the dodo of popular imagination. Despite the fact that Savery can hardly be considered an artistic giant, he is responsible for an image of truly iconic power and influence. And there are many artists of much greater stature who cannot make that claim! Whether his iconic vision is also an accurate portrait of the bird in life is something of which we cannot be sure.

Contemporary Writing

Just as there are pictures that were painted when dodos were actually alive, so too there are a few written descriptions that were penned by people who saw living dodos. Unfortunately, most of these reveal little about the creature. Seventeenth-century travelers were concerned with the edibility of animals and birds rather than with their habits, and most dodo accounts are of the "I went to Mauritius, I saw dodos, and I ate them" variety.

One particular account, though, stands out from all the others and is far more informative in its content. It was penned by an English courtier by the name of Thomas Herbert and relates to a visit he made to Mauritius during 1628. In his account of the island Herbert wrote:

> *First here only is generated the Dodo, which for shape and rareness may antagonize the Phoenix of Arabia: her body is round and fat, few weigh less than fifty pound. It is reputed more for wonder than for food, greasie stomackes may seeke after them, but to the delicate they are offensive and of no nourishment.*
>
> *Her visage darts forth melancholy, as sensible of Nature's injurie in framing so great a body to be guided with complementall wings, so small and impotent, that they serve only to prove her Bird.*
>
> *The halfe of her head is naked, seeming couered with a fine vaile, her bill is crooked downwards, in the midst is the thrill [nostril] from which part to the end 'tis of a light green, mixt with pale yellow tincture.*
>
> *Her eyes are small and like to Diamonds, round and rowling, her clothing downy feathers, her traine three small plumes, short and inproportionable, her legs suiting to her body, her pounces sharp, her appetite strong and greedy.*
>
> *Stones and iron are digested, which description will be better concieued in her representation.*

This famous passage is by far the most informative pen-portrait that survives from the era of the living dodo.

Opposite: The dodo with other Mauritius birds, drawn by Sir Thomas Herbert, 1628.

Sir Thomas Herbert, painted by an anonymous artist.

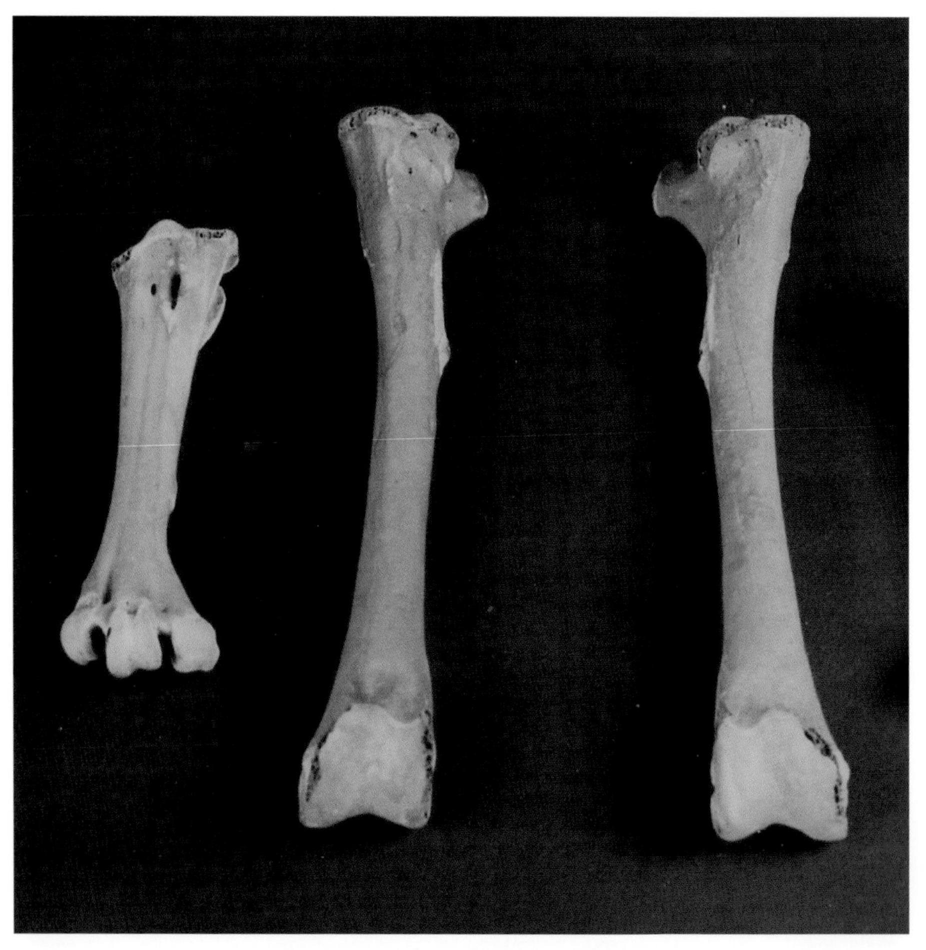

Dodo bones from the Mare aux Songes.

Bones and the Mare aux Songes

In the basements and backrooms of many of the world's museums there are quantities of dodo bones. Some museums possess many of them, others have just a few fragments. One of the curious things about these is the fact that almost all have a common origin. The vast majority of surviving bones are in a subfossil state and come from a small area of marshy land close to the south coast of Mauritius known as the Mare aux Songes.

A rather strange myth has grown up about this site, one that is difficult to account for. Several books and other publications have suggested that the swamp no longer exists and that it now lies buried beneath one of the runways of the present-day international airport. Just why this story has gained such popularity is difficult to say, for it is entirely false. The Mare aux Songes still exists and is situated a mile or two from the airport. Whether or not it still contains dodo bones is a matter for conjecture, however. It probably doesn't. In the years since its significance was first recognized the marsh has been investigated and excavated several times. It is therefore doubtful that it still houses interesting remains.

Just why this small area was so rich in dodo bones is something of a mystery. Obviously, dodos were once plentiful in the area, but whether the bones were deposited over a short period of time or whether they

The Mare aux Songes, Mauritius.

accumulated over centuries is not known. No research seems to have been conducted over the age of the bones, so it is not known whether they date from the era when man was visiting Mauritius or whether they come from a much earlier period. Carbon dating on a selection of the bones could, of course, help resolve this enigma.

The tale of the actual discovery of the bones is in itself a rather curious one. During the year 1865 a schoolmaster named George Clark, who happened to live nearby, heard that some tortoise bones had recently been found in the swampy ground. This fired his imagination and caused him to wonder if the Mare aux Songes might also hide bones of the dodo. The tantalizing area happened to be situated on the estate of a certain Monsieur de Bissy, who was the owner of a large sugar plantation known as Mon Trésor, Mon Désert. Clark persuaded de Bissy to allow him access to the land, but de Bissy was actually prepared to offer even more help. He allowed Clark to use the plantation workers (who at that time were treated little better than slaves) to help in the search. Clark directed these people to wade through the marsh (which, in most places, appears to have been approximately waist-deep), feeling with their toes for any irregularities on the bottom. In this rather bizarre manner many, many dodo bones were located and brought to the surface. Subsequent explorations revealed much more material, but the splendid cache of skeletal material seems to have been worked out long ago.

Mare aux Songes bones are easy to identify. They have a rich brown patina and are unlikely to be confused with bones from any other source. In fact, there are very few of the latter. Dodo bones from other Mauritian sites are rare, and considerable mystery and secrecy surrounds their discovery. The finders have, perhaps understandably, shown great reluctance to reveal the sites at which they have uncovered bones.

From all these bones, which have spread to museums all around the world, perhaps a dozen complete or almost complete skeletons have been constructed. These rare exhibits exist in several of the world's natural history museums.

Above and opposite:
Pictures of dodo skeletons
collected in the 19th century.

The Oxford Head

Sometime around the year 1650, a stuffed dodo arrived in the city of Oxford, England. No one knows how it got there or why. It is first mentioned in the catalog of the museum of a certain John Tradescant, once King Charles I's gardener. Tradescant was persuaded to bequeath the contents of his museum to the celebrated collector Elias Ashmole, and so the stuffed dodo passed into the collection of what is now the Ashmolean Museum. Here it stayed for almost a century, until, in January 1755, the trustees of the museum made their annual inspection of the stock.

One of Ashmole's original statutes (no. 8) reads:

> *That as any particular grows old and perishing the keeper may remove it Into one of the closets or other repository; and some other to be substituted.*

Under this statute the stuffed dodo was examined—and found wanting. It is easy enough to understand why. After 100 years of exposure to dust, dirt, and moths, the specimen must have presented a very sorry spectacle—and it was ordered out for immediate destruction. As far as "some other to be substituted"—none could be. No other stuffed dodo can be proved to have outlasted the Tradescant/Ashmole one. Presumably, the museum trustees had no real understanding of the importance of the object they had rejected. Certainly, they could not have realized that it was unique. None of the "stuffed dodos" that can be seen today in many of the world's natural history museums are "real." All are made up from chicken or goose feathers stuck onto a dodo-shaped frame.

A fake "stuffed" dodo.

The instruction to destroy the Oxford dodo was carried out, but not quite to the letter. Before the offending object was burned or thrown onto the dump (no one today knows

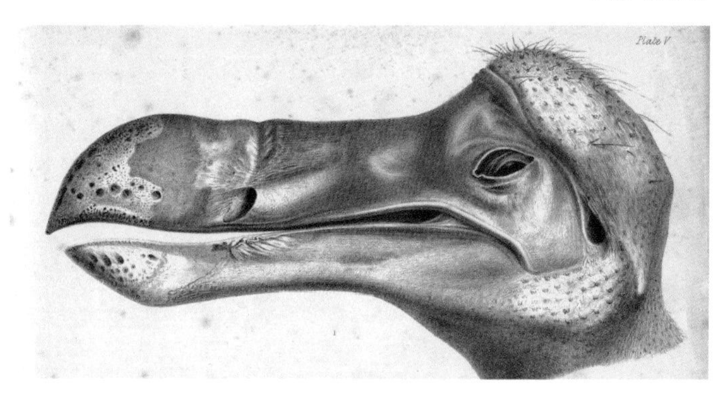

A 19th-century lithograph of the Oxford head.

how the disposal was effected), someone cut off the head and one of the feet, and these objects were carried back into the museum.

The head, in particular, has proved of immense importance and has served as the basis for several valuable anatomical studies. The foot never carried quite the same celebrity, since another one once existed (at the Natural History Museum, London) but now seems to have been lost.

The Oxford head and foot still exist, although they are no longer at the Ashmolean. It was felt that the University Museum of Zoology was a more appropriate location, and these famous relics were transferred long ago.

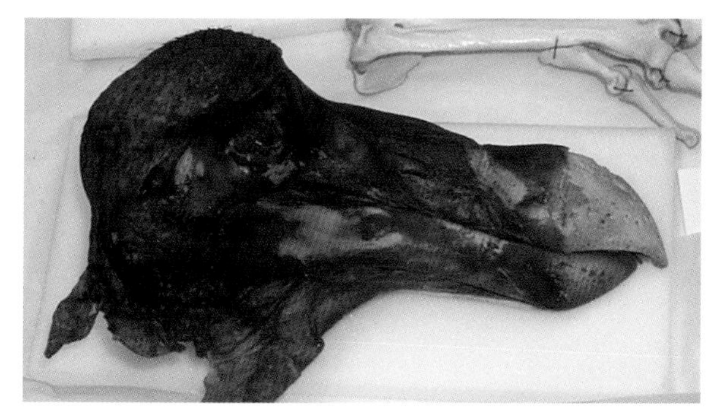

A modern photo of the Oxford head.

The Dodo and *Alice's Adventures in Wonderland*

The year 1865 is an important one in the history of the dodo, even though by this time no living dodo had walked the earth for around 200 years.

Before 1865 dodos were known only to academics and ardent students of zoology. After that year it was a household name. One single event was the cause of this change of fortune: the publication of Lewis Carroll's famous book *Alice's Adventures in Wonderland*.

The connection between the dodo and this book is a rather curious one. The character of Alice was modeled on a real little girl whom Lewis Carroll (real name Charles Lutwidge Dodgson) befriended. Both lived in Oxford, and this unlikely couple often went for long walks together around the city. Apparently, one of their favorite jaunts was a trip to the museum to see the celebrated remains of the dodo.

John Tenniel's original illustration for Alice's Adventures in Wonderland.

When Carroll decided to put his little friend Alice into his story, it was quite natural that he should also find a place for a dodo. Whether the simple inclusion of the bird into the narrative would have been enough to propel dodos to celebrity status cannot be said. But there was another factor. There can be little doubt that the immediate popularity of *Alice's Adventures in Wonderland* (as well as its lasting fame) owed much to the inventiveness of the illustrator. One of the great publishing coups of all time must surely be the selection of a certain John Tenniel to produce the pictures. The set of drawings that he made stand among the most evocative and memorable of all illustrations, and his picture of a dodo (based on the images of Roelandt Savery) leaning on a cane was certainly a launch pad for the dodo's surge to worldwide celebrity.

Within a year or two dodos were the rage. Almost everyone could recognize a dodo. People (particularly girls) were nicknamed "Dodo," and articles concerning the bird became commonplace in newspapers and magazines. The phrase "dead as a dodo" was coined and became widely used. The dodo had passed from vague myth to iconic status.

A more recent illustration by Angel Dominguez.

Dodo memorabilia:
A children's book, c. 1990;
a clock, c. 2000;
and a cigarette card, c. 1935.

Dodo Memorabilia

In keeping with its status as one of the great icons of extinction, dodo souvenirs of all kinds are hugely popular. Not only do these items appeal to the casual buyer, but there are also many seriously addicted "dodo collectors." Such people collect everything that has a dodo theme: dodo books, dodo models, dodo ashtrays, dodo teapots, dodo T-shirts, dodo clocks, dodo brooches, dodo anything! All manner of memorabilia is available to those who hoard such items. There are even perishables like candy, cookies, or hamburgers, each named after the dodo and carrying its image on the label or wrapper.

Devising and designing such commodities has become something of an industry, particularly on the island of Mauritius, where the image of the dodo can be seen in so many places. The island has a flourishing tourist industry, and it is natural that its shops and hotels should feature the country's greatest treasure. But, unlike the bird itself, dodo souvenirs are by no means unique to the island of Mauritius. With a high profile throughout the world and a visual appeal that seems universal, dodo goods can be found almost everywhere.

Different Kinds of Dodo?

Was there more than one kind of dodo? This is a question that has tantalized dodo enthusiasts for many years, and some of them have come to the conclusion that there were several. The reasons for such an opinion are varied, but it is mostly based on the discrepancies in the old paintings and in the writings of early travelers. Such discrepancies could, of course, be easily explained by inaccurate observation, poor painting skills, bad memory, or just plain tall-tale telling. On the other hand...

The most enthusiastically supported of these "other" dodos is one that has become known as the "white dodo of Réunion." Mauritius is one of three widely separated islands that together make up the group known as the Mascarenes, and Réunion is one of the others. There are some vague 17th-century reports that have led some ornithological historians to believe that a dodolike bird, white in color, once lived there, and this supposed creature has gained a kind of loose identity in ornithological literature. There are many good

The White Dodo of Réunion, *by an unknown artist, c. 1950.*

The Rodriguez Solitary, *by an unknown artist, c. 1950.*

reasons, however, for supposing that it never existed, and that the reports have been misunderstood. Unlike the dodo itself, not a single bone of this alleged creature has ever been found; nor is there any other kind of physical evidence to support its former existence. It would not be unreasonable to suppose that Réunion did indeed once have its own dodo, but so far not one piece of worthwhile evidence has been produced to show that it did.

There is no doubt, however, that a creature related to the dodo did once exist on the third of the Mascarene islands, the island of Rodriguez. Here, until the middle of the 18th century, there lived a bird known as the solitary. Although it was large in size it lacked the enormous beak and strange tail of the dodo and seems to have been an altogether plainer, less spectacular bird. There are two very clear written accounts that describe this bird in life, and also many bones have been found in the island's caves from which it has been possible to construct complete skeletons.

Perhaps one day, physical evidence of the former existence of others will be found. But no matter how many kinds of dodo there may once have been, there are none left now.

Further Reading

Fuller, E. *Dodo: From Extinction to Icon.* London and New York: HarperCollins, 2002.

_______. *Extinct Birds.* Oxford: Oxford University Press, 2000; and Ithaca: Cornell University Press, 2001.

Grihault, A. *The Dodo of Mauritius.* Mauritius: IPL Publications, 2002.

Hachisuka, M. *The Dodo and Kindred Birds.* London: Witherby, 1953.

Hume, J. "The Journal of the Flagship Gelderland–Dodo and Other Birds on Mauritius, 1601." *Archives of Natural History,* 30, no. 1 (2003).

Kitchener, A. "On the External Appearance of the Dodo." *Archives of Natural History* (1993), pp. 179-301.

Strickland, H. E., and A. G. Melville. *The Dodo and Its Kindred.* London: Benham, Reeves, 1848.

Ziswiler, V. *Der Dodo.* Zurich: Zoological Museum, 1996.

Some museums holding Dodo material

United States

Cambridge, Massachusetts: Museum of Comparative Zoology, Harvard University.
New York: American Museum of Natural History.

Great Britain

Cambridge: University Museum of Zoology.
Edinburgh: National Museums of Scotland.
London: The Natural History Museum.
Oxford: University Museum of Zoology.

Other Locations

Copenhagen: Zoological Museum, University of Copenhagen.
Durban, South Africa: Durban Museum.
Frankfurt: Senckenburg Museum of Natural History.
Paris: Natural History Museum.
Port Louis, Mauritius: Mauritius Institute.
Prague: Narodni Museum.
Vienna: Natural History Museum.